Theobald Smith

The Toxin of Diphtheria and Its Antitoxin

Theobald Smith

The Toxin of Diphtheria and Its Antitoxin

ISBN/EAN: 9783337816919

Printed in Europe, USA, Canada, Australia, Japan

Cover: Foto ©berggeist007 / pixelio.de

More available books at **www.hansebooks.com**

THE TOXIN OF DIPHTHERIA

AND ITS ANTITOXIN.

By THEOBALD SMITH, M.D.

George Fabyan Professor of Comparative Pathology in the Medical School of
Harvard University. In Charge of the Antitoxin Laboratory
of the State Board of Health.

Read at the Annual Meeting of the Massachusetts Medical Society,
June 7, 1898.

THE TOXIN OF DIPHTHERIA AND ITS ANTITOXIN.

EVER since Roux and Yersin demonstrated the fact that diphtheria bacilli act in the main through a soluble poison or toxin which they produce during multiplication, this toxin, as well as those of other pathogenic bacteria, has been the object of considerable attention on the part of experimental medicine. The discovery of the antitoxic properties of the blood of immunized animals by Behring and Kitasato in 1890 was a signal for the renewed study of the toxin for which a true antidote had now been found.

In accordance with their facilities and command of methods, experimenters have been exploiting the nature and action of toxins in different directions, some applying the methods of physiology and physiological chemistry, others those of pathological anatomy, still others those of experimental biology. Though we must acknowledge that we are only within the threshold of this work, it is difficult not to see what great progress has been made. Ten years ago a vague knowledge of the existence of bacterial poisons is all that we possessed. Of the existence of antitoxins there was not a hint. Since then a new horizon has come into view fully as great in its sweep as that which presented itself to the great discoverers in the domain of microörganisms. For while the latter revealed to us the swarm of enemies that beset us, the former unfolds before us the forces that the animal body holds in readiness for them.

An exhaustive review of the various additions to our stock of information upon the toxin of diphtheria would far exceed the time limit of this paper. I shall, therefore, restrict my remarks to a brief resumé of those facts and theories which appeal to the physician who is confronted by questions not in the laboratory but at the bedside.

The immediate objects of most of the investigations made in recent years may, for convenience, be grouped under three heads:

(1) A clearer insight into the nature of toxins and of their action upon the various tissues of the body and cells composing them, in order that diphtheria as well as other toxic diseases may be better understood.

(2) More information concerning the essential nature and specific action of antitoxin. And

(3) The improvement of the antitoxins and the cheapening of their production.

Before entering upon a discussion of these subjects it may be well to briefly review the established facts and some unestablished theories concerning the action of the diphtheria bacillus itself. While attention has been most exclusively focused upon the toxin it should be borne in mind that the bacillus is, after all, the exclusive manufacturer of this product, and that the character of the bacillus and its adaptation to the conditions found on mucous membranes determines from the beginning the gravity of the infection.

In the presence of suitable food substances, among which, in artificial cultures, the albumoses or so-called peptones seem to stand at the head, the diphtheria bacillus during its multiplication produces a poison or toxin which diffuses promptly from the cell body in which it presumably arises into the surrounding fluid. After removal of the bacilli from the culture fluid, this is capable of producing experimentally the various lesions including paralysis, ascribed to the diphtheria bacillus when it attacks the human subject. The bacillus, conveyed from one person to another, seems to vegetate almost exclusively upon exposed surfaces, primarily mucous membranes, secondarily wounded areas of the integument.

The hyperæmia and exudation which follow the earliest absorption of toxin, supply the bacillus with a suitable nidus in which to continue its multiplication and produce more toxin. In this way the diseased area spreads, the poison is manufactured and absorbed more abundantly, and constitutional symptoms indicated by fever and prostration appear. The microscope informs us that during the earliest local manifestations, the usual, scant, miscellaneous bacterial flora of the mucosa is quite suddenly replaced by a rich vegetation of the easily distinguishable diphtheria bacillus. Frequently no other bacteria are found in the culture tube. This vegetation continues for a few days, then gradually gives way to another flora of cocci and bacilli, and finally the normal condition is reëstablished. The disease process is probably checked in all cases by the appearance of antitoxin in the blood produced by certain tissues or cell groups under the stimulus of the diphtheria toxin. Now-a-days the antitoxin in the serum of the immunized horse is introduced to hasten and increase the resistance of the body.

Diphtheria bacilli vary slightly in their capacity to produce toxin in culture-fluids. A comparative study of over forty cultures of diphtheria bacilli from different towns of this State,* carried on under uniform conditions, showed that some bacilli produced from two to three times as much toxin in a given time as others. This quantitative production seems to remain indefinitely the same for each bacillus, so that the inference that there is some difference in the bacilli studied, cannot well be rejected. Just what bearing this difference would have upon the human subject we cannot easily determine, owing to the many

* Theobald Smith and E. L. Walker; Report of the State Board of Health of Massachusetts, 1897, p. 619.

variable conditions and the now prevalent antitoxin treatment.

Given the power to produce toxin, the important question arises as to the nature of the impulses which start the multiplication of diphtheria bacilli on mucous surfaces. The commonly accepted view is that the bacilli need only to reach the throat of a susceptible individual to multiply. But insusceptibility does not necessarily prevent multiplication, for this may go on for several months after the recovery of patients. If diphtheria bacilli could multiply on mucous membranes as soon as they reach them there would be many more healthy persons than are now found harboring diphtheria bacilli. I am inclined to believe that slight injury to the membrane, such as erosions, traumatism, hyperæmia followed by slight exudation due to exposure, and possibly certain undetermined chronic affections, favor and perhaps determine multiplication after infection in primarily susceptible individuals. It is highly probable that the special food needed by the diphtheria bacillus to produce toxins comes from the blood and becomes available through lesions of the mucous membrane. The injury may be of the slightest character and still afford a foothold for the bacilli, after which their multiplication with toxin production will lead to the spreading of the disease process.

This theory would help to account for the great increase of diphtheria in the colder season of the year, which increase is in part accounted for by the great overcrowding of living rooms in winter and the better opportunities for infection thereby offered. It would also account for the occasional discovery of bacilli in the throat a few days before the outbreak of the disease and harmonize with the experimentally demonstrated fact that toxins are harmless upon normal mucous membranes.*

* After the completion of this manuscript the work of Morax and Elmassian (Annal. de l'Institut Pasteur 1898, p. 210) on the action of the diphtheria

While infection will always play the dominant rôle in communicable diseases, the determining influence of accessory causes must not be lost sight of. These, I think, will form the great themes of study in the near future.

The persistence of diphtheria bacilli in the throat after the local and general disturbances have disappeared has been at first regarded as of little significance because the belief had been disseminated that such bacilli had lost the power to produce toxin. This, unfortunately, is not true. In the article quoted above, special attention was paid to this important subject. Even after a prolonged sojourn in the throat the toxin production of the bacilli was up to the average. Their harmlessness to the recovered patient may be referred to the presence of antitoxin in the blood, possibly also to the fact that the substances upon which the bacilli live in the normal throat may not yield that abundance of toxin furnished by the pathological exudates. Whatever the explanation may be, the potential danger of such cases as foci of the disease must be accepted in the light of recent researches.

Another subject of great importance, clinically, is the not infrequent occurrence of so-called septic forms of diphtheria, generally referred to the presence of adventitious pathogenic forms, such as streptococci and possibly of other little known bacteria. The source of the pathogenic forms is most consistently referred to some preëxisting lesion in the mouth or throat, such as catarrh, carious teeth, and lesions of the mucous membrane. Evidently the combina-

toxin on mucous membranes, came into my hands. These authors experimented upon the conjunctiva of rabbits. By instilling with the utmost caution a few drops of very strong toxin every 3 minutes for a period of 8 to 10 hours, characteristic diphtheritic changes of the conjunctival mucosa were produced. Less prolonged contact failed to cause any reaction. The nasal mucosa remained unaffected. This experiment simply shows that all tissues must give way eventually. If we assume that the mucosa in absorbing the toxin has the power of transforming it, this power will be broken down in time. Again, the instilling of a ready-made toxin of great strength does not mean that the diphtheria bacillus is in a position to manufacture it on the intact mucosa.

tions of bacteria with diphtheria bacilli may be many, and
the conditions favoring them very variable. No satisfactory light has been shed upon these mixed infections. We
know from almost daily experience that certain groups of
bacteria growing in nutritive fluids with diphtheria bacilli
favor the production of toxin materially, while other groups
decidedly interfere with it. The toxin produced under such
conditions is still diphtheria toxin, for it is neutralized by
antitoxin in the usual way. The frequently unsatisfactory
course of such cases may be due to the actual invasion of the
body by the accessory bacteria, or to the continued ex-
pansion of the disease process locally under their influence
whereby the diphtheria bacillus is favored in spite of antitoxic influences. Diphtheria antitoxin acts specifically upon
diphtheria toxin and on this only. Hence the interposition
of other pathogenic bacteria is likely to disturb the ex-
pected therapeutic action of the serum. The importance of
early treatment becomes manifest when we consider that the
septic bacteria may be at first favored by the diphtheria
bacillus, and then when once established return the favor
by supporting it.

It is also possible that non-specific inflammation of the
throat due to pyogenic bacteria may precede the infection
with diphtheria bacilli. The history of exposure and the
clinical course of the disease in the early stage, combined
with bacteriological examination of the throat, should bring
to light such cases if they actually occur.

The extensive bacteriological studies of mixed infections
made by Roux and Yersin, Martin, Barbier, Bernheim and
others, have not been able to establish certain somewhat
fanciful hypotheses of the action of streptococci upon diphtheria bacilli.* It would seem that the same processes

---

* In artificial mixed cultures the streptococci are probably unfavorable, the
staphylococci favorable to the production of diphtheria toxin. This I infer
from their general biological characters, as I have not made any direct experiments with them. Clinically, the streptococci, however, are regarded
as the most dangerous secondary invaders.

are at work as in ordinary wound infections. The diphtheria bacillus starts the lesion; the streptococci actually present take advantage of the wound to penetrate into the body. The greater the lesion and the more abundant the injured tissue and the exudation, the more favorable the conditions for the secondary invaders. The more virulent these are, as v. Dungern* states, the more severe the disease that follows their entry and multiplication. This element of virulence among the secondary invaders is probably the most potent : next would come the extent of the throat lesion upon which they are grafted. In this connection it may be interesting to note that Bonhoff† was able to produce in guinea-pigs a glomerulo-nephritis when cultures of diphtheria bacilli were injected, which had been prepared either by inoculating streptococcus cultures with diphtheria bacilli, and thus producing mixed cultures, or else by inoculating sterile filtrates of streptococcus cultures with diphtheria bacilli.

Experiment is unable to exploit many problems relating to bacteria pathogenic to man, and we must content ourselves in many respects with hypotheses adapted as closely as possible to truths already established. This difficulty obtains when we endeavor to seek information concerning the relative virulence of bacteria as distinguished from their toxicity. By virulence, I mean the power of bacilli to implant themselves upon living tissues, and resist bactericidal forces. It is evident that a virulent bacillus need not necessarily be a potent toxin producer. Virulent bacilli would include such as are readily infectious and which are likely to produce epidemics. Such epidemics need not be associated with a high mortality. The important factor is the spreading tendency of the disease. While we are able to compare the toxin production of different bacilli, the

* Beiträge zur allg. Pathol. XXI. p. 104.
† Hygien. Rundschau.

relative virulence is not determinable because there is no
species of animals which contract diphtheria spontaneously
and upon which experiments might be made.

Leaving now the bacillus itself, we will turn to the toxin.
This is demonstrable in culture fluids from which all bacilli
have been removed by its pathogenic properties.   Its action upon guinea-pigs is characteristic and the same quali-
tatively whatever culture may have produced the toxin.
The course of the experimental disease is determined by
the quantity of toxin administered.   Thus the dose which
is fatal to a guinea-pig in two and one-half days, when
doubled is, as a rule, fatal in one and one-half days.   With
the larger number of bacilli we have studied, about 0.08
cc. of the filtered culture fluid suffices to prove fatal to a
guinea-pig of 300 grams in three or four days.   A few
bacilli have been found which produce a toxin quantita-
tively ten times as strong as the average.   The lesions vary
in intensity according to the dose of toxin injected.   When
death ensues very rapidly within twenty-four hours, very
slight hemorrhagic changes are found at the place of injection.   When the disease lasts thirty-six hours, there is
more or less œdema associated with hemorrhage at the place
of injection, the lungs are œdematous, and are compressed
by a clear colorless effusion in the pleural sacs.   The ad-
renal bodies are deeply reddened.   When the disease lasts more
than two days, the pulmonary œdema and pleural effusion
are absent, and the local œdema is replaced by induration of
the tissues, sloughing and ulceration.   The whole process
seems to be started by an injury to the vascular walls
whereby œdema, thrombosis and hemorrhage, and the in-
evitable necrosis, follow one another in proper sequence.
The pulmonary œdema is also apparently the result of an
injury of the vascular walls by the toxin.   The lesions are
thus akin to those which appear on the mucous membrane
in the human subject.

Further than this the guinea-pig may show lesions of the nervous system manifested by disturbances of locomotion of greater or lesser severity and extent, involving the limbs and the muscles of respiration. The following table epitomizes the cases of nervous lesions among guinea-pigs observed during the past sixteen months in the laboratory among a total of 188, under observation from four weeks to several months after inoculation. These guinea-pigs received either toxin alone or else serum and toxin. The former group received a trifle less than the minimum fatal dose, the latter ten times that dose plus the antitoxin.

Of the forty-two guinea-pigs which survived the simple injection of toxin, there were represented all gradations of local changes from the merest œdema to necrosis and ulceration of the seat of injection. The table shows that four of these, or about ten per cent., became paralyzed. Of the 146 which survived the injection of the mixture of toxin and antitoxin, the local lesion was in most cases absent. Of these eight, or about five and one-half per cent., became paralyzed. In all cases of paralysis there was a local lesion, either very slight or more severe, indicating that the serum added to the toxin was not quite sufficient to neutralize the latter completely. The table furthermore shows that of these twelve cases, paralysis appeared in five cases in the third week, five cases in the fourth week, and three cases in the fifth week. In almost every instance the guinea-pig had been regarded as well and placed in a large box with other recovered animals. The paralytic lesions came on quite suddenly, usually over night. The animal was found moving with greater difficulty. The front limbs seemed to be attacked most frequently. The anterior half of the body rested upon the floor and the body was propelled by the hind feet. In one case prolonged friction with the floor of the box had produced bald spots over the sternum. When the hind limbs were affected, the body became unbal-

TABLE GIVING SUMMARY OF CASES OF PARALYSIS OBSERVED IN GUINEA-PIGS FOLLOWING THE SUBCUTANEOUS INJECTION OF DIPHTHERIA TOXIN.

NUMBER OF GUINEA-PIG.	WEIGHT IN GRAMS.	INJECTION OF		EFFECT (LOCAL).	DATE OF INJECTION.	DATE WHEN PARALYSIS FIRST NOTICED.	FINAL RESULT.
		TOXIN.	SERUM.				
519	340	Mixed Toxin, No. 44.	Horse IX.	Induration.	Feb. 10, 1897.	Mar. 11, 1897.	Recovered by March 24.
566	317	Toluol Toxin.	" X.	Very slight œdema.	Mar. 26, 1897.	Apr. 24, 1897.	Dies from paralysis of respiration Apr. 25-26.
597	339	" "	" "	" "	Apr. 20, 1897.	May 15, 1897.	Would probably die; chloroformed May 18.
604	300	" "	" III.	Very slight Induration and loss in weight.	Mar. 26, 1897.	Apr. 24, 1897.	Recovered May 7.
851	251	Toxin No. 5.	" "	Induration and bald area.	Dec. 21, 1897.	Jan. 14, 1898.	Recovered Jan. 28.
970	294	" " "	" XV.	Œdema followed by small slough.	Jan. 15, 1898.	Jan. 31, 1898.	Unable to move Feb. 5; chloroformed.
878	258	" " "	" IX.	Induration and slight slough.	Feb. 17, 1898.	Mar. 14, 1898.	Recovered March 23.
884	254	" " "	" XV.	Thin slough.	Feb. 17, 1898.	Mar. 13, 1898.	Dies March 17.
850	257	" Lb.	No serum.	Slough and ulcer.	Dec. 27, 1897.	Jan. 18, 1898.	Chloroformed.
899	277	" 3 Pf.	" "	Large slough.	Mar. 2, 1898.	Mar. 17, 1898.	Chloroformed March 21.
905	252	" No.5.	" "	Superficial slough.	Mar. 2, 1898.	Mar. 18, 1898.	Chloroformed; extensive paralysis.
901	277	" " 6.	" "	Slough.	Mar. 8, 1898.	Mar. 28, 1898.	Chloroformed; extensive paralysis.

anced, the animal swayed from side to side, in severe cases rolled over and gained its equilibrium with difficulty, or else remained lying down. In a few animals, respiration was affected. This became slower, sometimes labored, rarely almost convulsive. Such animals were generally incapacitated, lay on their sides and died soon or were chloroformed. In the milder forms recovery is complete in about two weeks from the date of the earliest paretic symptoms. The more severe cases die, probably of paralysis of the respiratory muscles. A consultation of the dates given in the table shows that cases of paralysis were not observed from May to December. Whether the warm season or the kind of toxin employed is to blame another year may help to explain. I am inclined to consider both elements involved. Cold weather certainly makes guinea-pigs less resistant to the toxin and the frequent occurrence of paralysis with toxin marked No. 5 points to this as in part to blame.

I have introduced these facts because they serve to illustrate once again the capacity of the toxin of diphtheria freed from the bacilli to produce lesions of the nervous system akin to those which appear in man.

While the toxin of diphtheria is thus recognizable only by what it does when injected into susceptible animals, we are not, therefore, at a loss to recognize its presence, though the process of recognition involves a great amount of labor in all investigations with it. The desire to know more definitely what the nature and composition of this substance is, is a very legitimate one, not only for theoretical but also for practical reasons. A definite chemical compound taking the place of the intangible pathogenic agent in our cultures would reduce the complex machinery needed in preparing antitoxin very greatly. Hence all toxins are now objects of research. Those of vegetable origin, such as ricin and abrin, those of animal origin, including the important snake poisons and the blood serum of various animals, and those

of bacterial origin, such as the toxins of diphtheria, tetanus and certain meat poisons, are receiving due attention ;
but the tendency to class these various poisons together and
to apply to all of them the slight and often vague information obtained from the study of any one of them, has left the
whole subject in a temporarily confused state. It is not
improbable that they belong to quite different groups, and
that each one must be carefully studied by itself. We
need only to compare the different conditions under which
the toxins of diphtheria and tetanus are produced,—the
one most advantageously with the most liberal supply of
oxygen, the other only when the activities of oxygen are
held in check,—to realize that these products may be
wholly unlike.

The toxin of diphtheria has been regarded as a ferment
or enzyme by Roux and Yersin, Sidney. Martin and others.
Later it was considered a toxalbumin by Brieger and Frän-
kel. More recently, however, Brieger recedes from that
position, and regards the poison as of non-proteid nature.
By an elaborate method he has concentrated and perhaps
purified the poison, without however obtaining enough for
analyses. He informs us that it is injured by acids and
oxidizing agents, but left intact by alkalies and reducing
agents. He found that after removing all the toxin from
the bodies of the bacilli, these still contained a poison,
probably derived from the cell substance, which produced
local necrosis and death in guinea-pigs. Antitoxin had no
power to neutralize this poison. Fortunately this does not
figure with any importance in human pathology, as the bacilli rarely penetrate into the body in any considerable num-
bers.

We know that the toxin of diphtheria is a more or less
unstable body. Heated in solution for several hours at
60 C., it loses almost wholly its property of producing disease. It likewise becomes weakened by exposure to air

and light. When dried it remains unchanged for a longer time. It is readily brought into this condition by adding ammonium sulphate to the filtered culture fluid. The peptones or albumoses are precipitated out and with them the toxin, which mixture is then dried in vacuo. Various other neutral salts, alcohol, and zinc chloride cause its precipitation mixed with other substances. It is also mechanically brought down when an insoluble phosphate is formed in the fluid containing it.

In the culture tube the formation of toxin is subjected to many vicissitudes. In the production of antitoxic serum of high potency, the most important condition to be fulfilled is the production of a very concentrated toxin; for the antitoxin, within certain limits, grows in strength with the amount of toxin the horse under treatment can endure. Hence, a study of the conditions under which the largest amount of toxin will appear in cultures has become the first to be entered upon. A discussion of the results of such a study upon which I myself have been engaged during the past three years would lead us too far into details which belong strictly to specialists, more particularly as they do not as yet throw much light upon the metabolism of the diphtheria bacillus or the immediate source and mode of formation of the toxin.

The toxin of diphtheria unfolds its power when introduced under the skin, directly into the blood or into the trachea. It is inert in the digestive tract. Once in the body it seems to disappear quite rapidly from the blood. According to Bomstein only one-eighth of the injected amount was found in the blood of rabbits after one hour. After five hours one-twentieth was detected. Nor was it traceable in the organs, in the contents of the intestines, or in the urine. The inference was drawn that it was destroyed by a chemical action of the tissues. Dzierzgowski believes that it is oxidized into antitoxin in the body ot highly immunized animals, a belief not generally shared at present.

The essential nature of the lesions produced by the toxin of diphtheria upon tissues and cells composing them has been the subject of many investigations. To obtain a simple morphological expression of the effect of this poison is an object worth striving for if it is indeed ever realizable. The lesions observed in the throat of patients and in the subcutis of guinea-pigs after an injection of the toxin do not reveal anything specific, and are readily ranged under the head of inflammation. The lesions observed in the cardiac muscle in the experimentally induced disease have been regarded by Comba, Mollard and Regaut and others as primarily affecting the muscle fibres. Interstitial lesions are slight and purely secondary. When we come to the nervous system to which the observed paralyses early drew the attention of histologists, a number of non-specific changes are recognizable. In the central nervous system, degenerative changes in the cells of the anterior horns have been described recently by Murajeff as following all injections of toxin whether leading to paralyses or not, but similar changes have been found by Goldscheider and Flatau for tetanus, by Kempner and Pollack for the poison of botulismus (meat-poison), and by Kossel for the poison of bloodserum from eels. The work of studying the lesions of the nervous system due to toxins has been begun in earnest and in due time this subject will become clarified. At present we are not in position to deduce from such work any simple conclusions. The peripheral nerves have also received their share of attention and the paralyses referred to a neuritis as well as to degenerative changes. Quite recently Dr. J. J. Thomas* has published interesting studies upon the degenerative changes observed in the pneumogastric nerve in those cases in which heart complications followed diphtheria.

---

* Boston Med. and Surg. Journal. 1898.

To these miscellaneous lesions associated with diphtheria we must add the diffuse and focal cell necroses in various organs observed by Oertel, Welch and Flexner, Barbacci, and others, and the appearance, especially in the kidneys, of peculiar cells by Councilman, denominated plasma cells, and probably derived, according to him, from lymphoid elements. In view of these various apparently unrelated phenomena, we cannot ascribe to the toxin of diptheria any specific easily recognizable effect upon the body, but must be content at present with assuming that the toxic activity embraces tissues of different orders and with variously differentiated functions, or else that it affects injuriously some one tissue incorporated with the various specialized tissues. This one tissue seems to be that composing the vascular system, and I am inclined to look upon injury to the vascular walls, especially those of the capillaries, as perhaps the *fons et origo* of the whole set of pathological changes. When the toxin is injected in very large quantities, the effect, as Barbacci* states, passes beyond the domain of the vascular system and is manifest in other tissues as well. This view does not, in any way, stand opposed to the one that diphtheria toxin is essentially a cell poison with a special affinity for the nucleus.

Moving parallel with these investigations are those forced upon us by the discovery and applicability of diphtheria antitoxin as a curative agent. The importance of accurately gauging the neutralizing power of antitoxin upon the toxin has led to much study upon the immediate relation of the toxin of diphtheria to its antitoxin. Considerable light has been shed upon the whole subject from this, the biological direction.

It will be remembered that two views of the action of antitoxins have been current. One assumes that the antitoxins stimulate the cells of the body whereby the latter

* Centralb. f. allg. Pathol. 1896. No. 8.

are put into a condition to neutralize or destroy the toxins. The action of antitoxins upon toxins is, according to this theory, indirect. The other theory claims a direct action of antitoxin upon toxins. The actual demonstration of this direct action has been attempted by various observers, but, owing to the difficulties in the way of experiments of this nature, the results are conflicting. The direct action upon toxins of their specific antitoxins has, however, been demonstrated for two other toxins, one of vegetable origin, the toxalbumin of the castor bean, ricin; and one of animal origin, the toxin in the bloodserum of the *Murœnidæ* or eels.

Ricin, which was first studied by Ehrlich, has the power to produce an agglutination of red corpuscles both in the body and in extravascular blood. In the body the clots of red corpuscles produced by minute doses of ricin lead to more or less extensive sloughing of the skin if the animal survives. Ehrlich has demonstrated the important fact that susceptible animals can be made immune to large doses of this poison by treating them with gradually increasing doses. The blood and the milk of these immunized animals possess protective powers; are, in other words, antitoxic.

Recently he has shown that the antitoxin acts directly upon the toxin in the test-tube. When ricin is added to blood, kept fluid by citrate of soda, it lumps the red corpuscles together. The masses thus formed settle down. If serum from an immunized animal be added to the blood, this clumping does not take place upon the addition of ricin. The phenomena take place under definite quantitative laws. A given amount of a certain serum neutralizes a given amount of ricin and no more.

Kossel has extended this proof by experimenting with the poisonous blood serum of eels. This serum, of which 0.1 cc. is sufficient to kill a rabbit weighing a kilogram in

three to four minutes, produces, when administered in
smaller doses, emaciation, partial paralysis of the ex-
tremities, and death after several days. It dissolves red
corpuscles. Rabbits can be immunized by injecting very
minute doses. When eel serum is added to defibrinated,
diluted blood, the coloring matter is promptly extracted
from the corpuscles. If, however, some serum from an
immunized rabbit be added to the blood beforehand, the
corpuscles remain intact. Here also the quantity of immune serum needed is directly proportional to the quantity
of eel serum used. These experiments are destined to
strengthen the hypothesis that antitoxins act upon or combine with their respective toxins and thereby render them
harmless. They do not, however, account for the whole
process of toxin immunity into which other factors probably
enter. Much more persistent experimentation will be
needed to clear up the obscurity that still clings to the
marvellous action of antitoxin.

To explain more consistently the various phenomena of
antitoxic action, Ehrlich has formulated a very ingenious
theory relating both to the mode of production of antitoxins in the body and their action upon toxins. This
theory, which has been termed latterly the lateral-chain
theory of immunity (*Seitenkettentheorie*), seems to have
not only stimulated researches but to have gained strength
through such researches. In a recent paper* I made a very
brief statement of this theory, which for the sake of making
subsequent statements more intelligible I will repeat here.

If we regard the cell substance as composed of complex
molecules with lateral chains, the toxin is assumed to unite
with one or more of these lateral chains, thereby incapaci-
tating the cell. This union is due to an affinity of the
toxin for the cell substance, which affinity expresses itself

---

* The Toxin and Antitoxin of Tetanus. Boston Medical and Surgical
Journal. 1898. p. 292.

to us clinically as susceptibility. If there were no affinity
according to this theory, there would be no susceptibility,
no disease. If only very minute doses of toxin are introduced so as not to incapacitate the cell substance too much,
that part lost by union with the toxin is reproduced by the
cell. The repeated injection of gradually increasing doses
of poison, if cautiously done, stimulates the cell, so to
speak, to an ever-increasing production of that part of
itself to which the toxin becomes attached. This over-
production of a specific cell substance leads finally to a
shedding into the blood stream where the shed material
appears as an antitoxin ready to combine with any toxin
for which it possesses definite affinities. This substance,
as we know, is effective not only in the blood of the producer but in the blood of other beings into which it is introduced.

This hypothesis has been received with much interest and
its general applicability will no doubt be thoroughly tested.
Already Wasserman* has shown that in harmony with it
the substance of the normal spinal cord and of the brain
possesses well marked affinities for the tetanus toxin. This
affinity is shown by the fact that a given amount of nervous
tissue neutralizes a certain quantity of tetanus toxin. Thus
three times the fatal dose of toxin, mixed with one-third of
the finely crushed spinal cord of a guinea-pig suspended in
salt solution, produces no effect when injected. The toxin,
in other words, has been, in some way, neutralized by the
tissues of the cord. Similar experiments carried out by
Bomstein to determine those cell territories having a special
affinity for the diphtheria toxin led to no satisfactory results.

Concerning the nature of antitoxins we are as much in
the dark as with toxins. We know, however, that they
are much more stable substances than toxins. The anti-

* Berl. klin. Wochenschr. 1898. p. 209.

toxic power of blood serum from immunized animals is fairly permanent even after many months. It resists freezing and even exposure to direct sunlight for a short time. It can endure a higher temperature than the toxin without losing its efficacy. Various conjectures have been made as to its character, some claiming that it is a very specific substance, others that the antitoxic power results from a change in certain regular constituents of the blood. It is precipitable by various substances which precipitate the albumins and globulins of the blood and seems to go with them when the serum is separated in layers of different densities by freezing. Smirnow has made the claim that antitoxins can be produced from toxins by electrolysis, but his claims, at least so far as the production of an effective substance goes, seem to grow more shadowy.

When antitoxin is injected into animals, it begins to disappear according to Behring on the fourth day, appearing then in the milk and the urine. The blood contains the greater amount and it would seem that most of it, at least, remains in the blood until it disappears from the body. In horses which have been actively immunized with toxins, Dzierzgowski found the largest amount of antitoxin in the blood and serous fluids, including the œdema formed where the toxin is injected. In the organs there is comparatively little to be detected. Bomstein found that if enough antitoxin be injected into dogs to supply seven units per cc. of blood, three units were found on the third day, 1.5 on the the sixth, 0.3 on the fourteenth, and none on the eighteenth day; the organs did not absorb it, as they contained very little; similarly very little was found in the urine. These facts agree very well with the restricted period of passive immunity induced by antitoxin.

The occasionally observed toxic effect of antitoxic serum must be ascribed to a peculiar personal idiosyncrasy manifested in the presence of horse's serum. Likewise the

rashes and occasional joint affections arc referable to the
serum and not to the antitoxin. We are, however, fortu-
nate in having horse's serum to use, since Uhlenhuth has
recently shown that compared with the serum from a number of other mammals it is by far the least toxic of them
all.

In presenting you with a large amount of miscellaneous,
somewhat conflicting and but feebly coherent information
upon the toxin of diphtheria and its antitoxin, I shall feel
gratified if it will simply convey to you the conviction that
toxins and antitoxins arc realities, entities, whose obscure
nature and action need not stand in the way of our belief
that we have reached the right path in studying them and
that we need but go on to gain, if not wider and deeper
information, at least better practical results in the combat-
ing of toxic disease.

However, we still hear voices occasionally raised in medical journals against the diphtheria bacillus as the chief
cause of the disease whose name it bears. This is not sur-
prising. Clinical experience is so subject to chance and at
the mercy of unknown factors that discussions of this nature arc likely to appear as long as medical science shall be
cultivated. The unambiguous results obtained by experimentation in the laboratory, though often too rashly ap-
plied, become more and more the guiding star of the young
physician as he enters upon his profession. Laboratory
work will give not only a steadiness of the judgment but
also an abiding sense of the circumscription of our knowledge and of the need for its continual rectification. If the
laboratory did nothing else it would amply repay to the
profession and its clients the costs of maintenance.

www.ingramcontent.com/pod-product-compliance
Lightning Source LLC
Chambersburg PA
CBHW022034190326
41519CB00010B/1714